WALKING IN THE FOOTSTEPS OF THE BRAVE

LAWANDA D. WARTHEN
ILLUSTRATED BY TAILLEFER LONG

WALKING IN THE FOOTSTEPS OF THE BRAVE
Second Edition, Copyright © 2021
Originally Published in 2012

ISBN: 978-1-7366170-0-7
Library of Congress Control Number: 2021902325

Copyright © 2021 Lawanda Denise Warthen
All rights reserved.

No part of this book may be reproduced, stored in retrieval systems, or transmitted in any form, by any means, including mechanical, electronic, photocopying, recording or otherwise, without prior written permission of the author.

Illustrations and Design by Taillefer Long
info@IlluminatedStories.com

DEDICATION

Freedom is not free – it comes at a price.
This book was written to honor the men and women who made the ultimate sacrifice so that we can all enjoy the freedoms we have today.

To my grandfather, Staff Sergeant Willie L. Prater. Thank you for serving during the Battle of the Bulge and paving the way.

To Mom and Dad, with much love and respect.

To my brother and sister, Troy Williams and Willette Odom, for their trustworthiness and support.

To my nieces and great-nieces, Kiah, Danayla, Zarah and Loriana…all I do is for you.

- LDW

A MESSAGE FROM THE AUTHOR

I am an African American woman who grew up in a society still recovering from the effects of segregation, I remember the effects of racial discrimination on a societal level. It wasn't until a visit to the Holocaust Museum in Washington with my two nieces, that I realized that the younger generation of African Americans has very little understanding of the suffering endured by both racial and ethnic minorities before and during World War II.

When I was stationed at an Army hospital in Germany in 2005, I learned about the "Wereth 11," an African American U.S. Army battalion that bravely fought against the Nazi SS at the Battle of the Bulge during World War II. I was surprised that I had never heard of such a significant historical event. As I began to research, I crossed paths with Joseph Small, an American filmmaker. Joseph was planning to film a documentary about the townspeople of Wereth, Belgium, and how they had given our soldiers food and shelter before they were captured and killed. We both participated in the dedication ceremony for the monument of the townspeople and soldiers.

This monument stands today as the only memorial in Europe that recognizes the service and sacrifice of African American soldiers who fought for our country and died during WWII. It is also dedicated to the town of Wereth and is maintained in perpetuity by U.S. Wereth Memorial Foundation Committee and The ROCKS, Inc., the largest professional military officers' organization where African Americans represent the membership majority.

Believing that great human sacrifice and historical milestones must be remembered so humanity will learn from them and not make the same mistakes, I decided to write this book. I hope it encourages all parents to teach their children about tolerance and the sacrifice made for peace and freedom.

-Lawanda D. Warthen

WALKING IN THE FOOTSTEPS OF THE BRAVE.

FLASHBACK

Over 65 years ago, the 333rd Field Artillery soldiers spent six months in combat supporting the 2nd Infantry Division and the 7th Corps, holding the front line against German troops.

I WANT TO WALK IN THE FOOTSTEPS OF THE BRAVE.

FLASHBACK
Before the war, people lived in peace. They lived in nice houses, had families, and took care of each other.

AMERICANS WERE ASKED TO JOIN THE ARMED FORCES TO HELP WIN THE WAR.

FLASHBACK

The spirit of these eleven brave men, who lost their lives at the hands of NAZI SS officers, is commemorated in the tiny village of Wereth, in Belgium.

ELEVEN BLACK MEN JOINED THE ARMY.

William Edward Pritchett – Alabama
James A. Stewart - West Virginia
Thomas J. Forte -Mississippi
Mager Bradley – Mississippi
George Davis - Alabama
James L. Leatherwood - Mississippi
George W. Moten - Texas
Due W. Turner - Arkansas
Curtis Adams – South Carolina
Robert Green - Mississippi
Nathaniel Moss -Texas

GOD GAVE THEM HIS STRENGTH AND COURAGE TO FIGHT FOR OUR COUNTRY.

"The path of the righteous man is beset on all sides by the inequities of the selfish and the tyranny of evil men. Blessed is he, who in the name of charity and goodwill shepherds the weak through the valley of darkness, for he is truly his brother's keeper and the finder of lost children."

- Ezekiel 25:17

THE SOLDIERS WERE FROM THE SOUTH. THEY WERE YOUNG.

FLASHBACK

These 11 men, looking for food and shelter, wandered to Wereth, where they stumbled upon a farm. The owner, Matthais Langer, was gracious enough to open his home to them, offering hospitality.

However, the NAZI SS received a tip that Americans were seeking shelter at the Langer's home. The Germans took the American soldiers from Langer's house, marching them down a hill to their death.

**I WANT TO WALK
IN THE FOOTSTEPS
OF THE BRAVE.**

THE MEN LEFT THEIR WIVES.

THEY HAD TO LEAVE THEIR CHILDREN AND FAMILIES BEHIND.

THIS SPECIAL GROUP OF AFRICAN AMERICAN SOLDIERS FOUGHT TOGETHER TO PROTECT US FROM ENEMIES.

THEY FOUGHT IN THE COLD, RAIN, AND SNOW. THEY FOUGHT DAY AND NIGHT. SOMETIMES THEY HAD TO LIVE ON THE BATTLEFIELD.

THESE MEN ENDURED MORE THAN MOST OF US CAN IMAGINE.

WALKING IN THE FOOTSTEPS OF THE BRAVE.

FLASHBACK

Although the Langer family erected a small cross with the names of the slain soldiers, their story was not well documented.

The memorial and the town of Wereth were not listed in any guides or maps referring to the Battle of the Bulge.

THEY WERE NOT TREATED FAIRLY, BUT THEY STOOD TALL.

FLASHBACK

In 2001, the townspeople of Wereth began raising money to build a bigger memorial, so that the world would know about these eleven brave men.

The new memorial was dedicated on May 23, 2004. In addition to the monuments built by the residents, the U.S. Veterans Chapter of the Battle of the Bulge added a plaque.

THEY TOOK CARE OF EACH OTHER WHEN THEY WERE HURT.

THEY FOUGHT AND DIED LIKE MEN. THEY GAVE THEIR LIVES.

FLASHBACK

The ROCKS Inc. European Officers Chapter, U.S. active-duty, reserve, retired Soldiers, family members and local town people in Germany and Belgium participated in a ceremony commemorating the men known as the "Wereth 11."

THEY ARE THE BRAVE ONES. DO YOU SEE WHERE THEY WALKED?

THE STORY OF THESE BRAVE SOLDIERS FINALLY MADE IT HOME.

**WE HONOR THESE
11 BRAVE MEN WHO
SACRIFICED THEIR LIVES
FOR THEIR COUNTRY
AND FOR JUSTICE.**

TODAY, WE CELEBRATE THE LIVES OF THESE HEROIC MEN.

THE MEMORY OF THE SOLDIERS IS STILL ALIVE.

ABOUT THE AUTHOR

Lawanda Denise Warthen grew up in West Palm Beach, Florida. She is a graduate of Walden University, and holds a Doctorate in Healthcare Administration.

She resides in Woodbridge, Virginia, and works as the Director, Public Affairs, Army Medicine, Falls Church, Virginia – Defense Health Headquarters. She enjoys reading and traveling with family and friends.

www.ingramcontent.com/pod-product-compliance
Lightning Source LLC
Chambersburg PA
CBHW052304200426
43209CB00069B/1897/J